This planner belongs to :

_____

2025

January

S	M	T	W	T	F	S
			1	2	3	4
5	6	7	8	9	10	11
12	13	14	15	16	17	18
19	20	21	22	23	24	25
26	27	28	29	30	31	

February

S	M	T	W	T	F	S
						1
2	3	4	5	6	7	8
9	10	11	12	13	14	15
16	17	18	19	20	21	22
23	24	25	26	27	28	

March

S	M	T	W	T	F	S
						1
2	3	4	5	6	7	8
9	10	11	12	13	14	15
16	17	18	19	20	21	22
23	24	25	26	27	28	29
30	31					

April

S	M	T	W	T	F	S
		1	2	3	4	5
6	7	8	9	10	11	12
13	14	15	16	17	18	19
20	21	22	23	24	25	26
27	28	29	30			

May

S	M	T	W	T	F	S
				1	2	3
4	5	6	7	8	9	10
11	12	13	14	15	16	17
18	19	20	21	22	23	24
25	26	27	28	29	30	31

June

S	M	T	W	T	F	S
1	2	3	4	5	6	7
8	9	10	11	12	13	14
15	16	17	18	19	20	21
22	23	24	25	26	27	28
29	30					

July

S	M	T	W	T	F	S
		1	2	3	4	5
6	7	8	9	10	11	12
13	14	15	16	17	18	19
20	21	22	23	24	25	26
27	28	29	30	31		

August

S	M	T	W	T	F	S
					1	2
3	4	5	6	7	8	9
10	11	12	13	14	15	16
17	18	19	20	21	22	23
24	25	26	27	28	29	30
31						

September

S	M	T	W	T	F	S
	1	2	3	4	5	6
7	8	9	10	11	12	13
14	15	16	17	18	19	20
21	22	23	24	25	26	27
28	29	30				

October

S	M	T	W	T	F	S
			1	2	3	4
5	6	7	8	9	10	11
12	13	14	15	16	17	18
19	20	21	22	23	24	25
26	27	28	29	30	31	

November

S	M	T	W	T	F	S
						1
2	3	4	5	6	7	8
9	10	11	12	13	14	15
16	17	18	19	20	21	22
23	24	25	26	27	28	29
30						

December

S	M	T	W	T	F	S
	1	2	3	4	5	6
7	8	9	10	11	12	13
14	15	16	17	18	19	20
21	22	23	24	25	26	27
28	29	30	31			

Year in Pixels

	J	F	M	A	M	J	J	A	S	O	N	D
1.												
2.												
3.												
4.												
5.												
6.												
7.												
8.												
9.												
10.												
11.												
12.												
13.												
14.												
15.												
16.												
17.												
18.												
19.												
20.												
21.												
22.												
23.												
24.												
25.												
26.												
27.												
28.												
29.												
30.												
31.												

Color Codes

Notes

January

2025

MONDAY	TUESDAY	WEDNESDAY	THURSDAY
		1	2
6	7	8	9
13	14	15	16
20	21	22	23
27	28	29	30

January

2025

FRIDAY	SATURDAY	SUNDAY	NOTES
3	4	5	○
			○
			○
			○
			○
10	11	12	○
			○
			○
			○
			○
17	18	19	○
			○
			○
			○
24	25	26	○
			○
			○
			○
			○
31			NOTES

February 2025

MONDAY	TUESDAY	WEDNESDAY	THURSDAY
3	4	5	6
10	11	12	13
17	18	19	20
24	25	26	27

February

2025

FRIDAY	SATURDAY	SUNDAY	NOTES
	1	2	○
			○
			○
			○
			○
7	8	9	○
			○
			○
			○
14	15	16	○
			○
			○
			○
			○
21	22	23	○
			○
			○
			○
			○
28			NOTES

March

MONDAY	TUESDAY	WEDNESDAY	THURSDAY
		1	2
6	7	8	9
13	14	15	16
20	21	22	23
27	28	29	30

March 2025

FRIDAY	SATURDAY	SUNDAY	NOTES
3	4	5	○
			○
			○
			○
			○
10	11	12	○
			○
			○
			○
			○
17	18	19	○
			○
			○
			○
			○
24	25	26	○
			○
			○
			○
			○
31			NOTES

April

2025

MONDAY	TUESDAY	WEDNESDAY	THURSDAY
	1	2	3
7	8	9	10
14	15	16	17
21	22	23	24
28	29	30	

April

2025

FRIDAY	SATURDAY	SUNDAY	NOTES
4	5	6	○
			○
			○
			○
			○
11	12	13	○
			○
			○
			○
18	19	20	○
			○
			○
			○
			○
25	26	27	○
			○
			○
			○
			○
			NOTES

May

MONDAY	TUESDAY	WEDNESDAY	THURSDAY
			1
5	6	7	8
12	13	14	15
19	20	21	22
26	27	28	29

May 2025

FRIDAY	SATURDAY	SUNDAY	NOTES
2	3	4	○
			○
			○
			○
9	10	11	○
			○
			○
			○
16	17	18	○
			○
			○
			○
23	24	25	○
			○
			○
			○
30	31		NOTES

June

2025

MONDAY	TUESDAY	WEDNESDAY	THURSDAY
2	3	4	5
9	10	11	12
16	17	18	19
23	24	25	26

June

FRIDAY	SATURDAY	SUNDAY	NOTES
		1	○
			○
			○
			○
			○
6	7	8	○
			○
			○
			○
13	14	15	○
			○
			○
			○
			○
20	21	22	○
			○
			○
			○
			○
27	28	29	30

July

MONDAY	TUESDAY	WEDNESDAY	THURSDAY
	1	2	3
7	8	9	10
14	15	16	17
21	22	23	24
28	29	30	31

July

2025

FRIDAY	SATURDAY	SUNDAY	NOTES
4	5	6	○
			○
			○
			○
			○
11	12	13	○
			○
			○
			○
18	19	20	○
			○
			○
			○
			○
25	26	27	○
			○
			○
			○
			○
			NOTES

August

MONDAY	TUESDAY	WEDNESDAY	THURSDAY
4	5	6	7
11	12	13	14
18	19	20	21
25	26	27	28

August 2025

FRIDAY	SATURDAY	SUNDAY	NOTES
1	2	3	○
			○
			○
			○
			○
8	9	10	○
			○
			○
			○
			○
15	16	17	○
			○
			○
			○
			○
22	23	24	NOTES
29	30	31	

September

MONDAY	TUESDAY	WEDNESDAY	THURSDAY
1	2	3	4
8	9	10	11
15	16	17	18
22	23	24	25
29	30		

September 2025

FRIDAY	SATURDAY	SUNDAY	NOTES
5	6	7	○
			○
			○
			○
			○
12	13	14	○
			○
			○
			○
19	20	21	○
			○
			○
			○
			○
26	27	28	○
			○
			○
			○
			○
			NOTES

October

MONDAY	TUESDAY	WEDNESDAY	THURSDAY
		1	2
6	7	8	9
13	14	15	16
20	21	22	23
27	28	29	30

October 2025

FRIDAY	SATURDAY	SUNDAY	NOTES
3	4	5	○
			○
			○
			○
			○
10	11	12	○
			○
			○
			○
17	18	19	○
			○
			○
			○
			○
24	25	26	○
			○
			○
			○
			○
31			NOTES

November 2025

MONDAY	TUESDAY	WEDNESDAY	THURSDAY
3	4	5	6
10	11	12	13
17	18	19	20
24	25	26	27

November 2025

FRIDAY	SATURDAY	SUNDAY	NOTES
	1	2	○
			○
			○
			○
			○
7	8	9	○
			○
			○
			○
14	15	16	○
			○
			○
			○
			○
21	22	23	○
			○
			○
			○
			○
28	29	30	NOTES

December 2025

THURSDAY	TUESDAY	WEDNESDAY	THURSDAY
1	2	3	4
8	9	10	11
15	16	17	18
22	23	24	25
29	30	31	

December 2025

FRIDAY	SATURDAY	SUNDAY	NOTES
5	6	7	○
			○
			○
			○
			○
12	13	14	○
			○
			○
			○
			○
19	20	21	○
			○
			○
			○
26	27	28	○
			○
			○
			○
			○
			NOTES

December
2024

01 SUNDAY

02 MONDAY

03 TUESDAY

04 WEDNESDAY

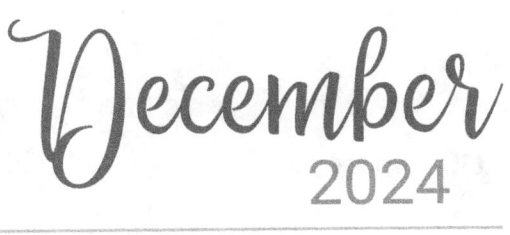

December
2024

05 THURSDAY
- ○
- ○
- ○
- ○
- ○
- ○
- ○
- ○

06 FRIDAY
- ○
- ○
- ○
- ○
- ○
- ○
- ○
- ○

07 SATURDAY
- ○
- ○
- ○
- ○
- ○
- ○
- ○
- ○

08 SUNDAY
- ○
- ○
- ○
- ○
- ○

09 MONDAY

10 TUESDAY

11 WEDNESDAY

12 THURSDAY

December
2024

13 FRIDAY
- ○ _____
- ○ _____
- ○ _____
- ○ _____
- ○ _____
- ○ _____
- ○ _____
- ○ _____

14 SATURDAY
- ○ _____
- ○ _____
- ○ _____
- ○ _____
- ○ _____
- ○ _____
- ○ _____
- ○ _____

15 SUNDAY
- ○ _____
- ○ _____
- ○ _____
- ○ _____
- ○ _____
- ○ _____
- ○ _____
- ○ _____

16 MONDAY
- ○ _____
- ○ _____
- ○ _____
- ○ _____
- ○ _____

December
2024

17 TUESDAY

○ _____
○ _____
○ _____
○ _____
○ _____
○ _____
○ _____
○ _____

18 WEDNESDAY

○ _____
○ _____
○ _____
○ _____
○ _____
○ _____
○ _____
○ _____

19 THURSDAY

○ _____
○ _____
○ _____
○ _____
○ _____
○ _____
○ _____
○ _____

20 FRIDAY

○ _____
○ _____
○ _____
○ _____

21 SATURDAY

○ _____
○ _____
○ _____
○ _____
○ _____
○ _____
○ _____
○ _____

22 SUNDAY

○ _____
○ _____
○ _____
○ _____
○ _____
○ _____
○ _____
○ _____

23 MONDAY

○ _____
○ _____
○ _____
○ _____
○ _____
○ _____
○ _____
○ _____

24 TUESDAY

○ _____
○ _____
○ _____
○ _____
○ _____

December
2024

25 WEDNESDAY

26 THURSDAY

27 FRIDAY

28 SATURDAY

December
2024

29 SUNDAY

○ _____
○ _____
○ _____
○ _____
○ _____
○ _____
○ _____
○ _____

30 MONDAY

○ _____
○ _____
○ _____
○ _____
○ _____
○ _____
○ _____
○ _____

31 TUESDAY

○ _____
○ _____
○ _____
○ _____
○ _____
○ _____
○ _____
○ _____

NOTES

_____
_____
_____
_____

January
2025

01 WEDNESDAY

02 THURSDAY

03 FRIDAY

04 SATURDAY

January
2025

05 SUNDAY

○ _____
○ _____
○ _____
○ _____
○ _____
○ _____
○ _____
○ _____

06 MONDAY

○ _____
○ _____
○ _____
○ _____
○ _____
○ _____
○ _____
○ _____

07 TUESDAY

○ _____
○ _____
○ _____
○ _____
○ _____
○ _____
○ _____
○ _____

08 WEDNESDAY

○ _____
○ _____
○ _____
○ _____
○ _____

09 THURSDAY
○ _____
○ _____
○ _____
○ _____
○ _____
○ _____
○ _____
○ _____

10 FRIDAY
○ _____
○ _____
○ _____
○ _____
○ _____
○ _____
○ _____
○ _____

11 SATURDAY
○ _____
○ _____
○ _____
○ _____
○ _____
○ _____
○ _____
○ _____

12 SUNDAY
○ _____
○ _____
○ _____
○ _____
○ _____

January
2025

13 MONDAY

14 TUESDAY

15 WEDNESDAY

16 THURSDAY

17 FRIDAY

- ○
- ○
- ○
- ○
- ○
- ○
- ○
- ○

18 SATURDAY

- ○
- ○
- ○
- ○
- ○
- ○
- ○
- ○

19 SUNDAY

- ○
- ○
- ○
- ○
- ○
- ○
- ○
- ○

20 MONDAY

- ○
- ○
- ○
- ○

January
2025

21 TUESDAY

○ _____
○ _____
○ _____
○ _____
○ _____
○ _____
○ _____
○ _____

22 WEDNESDAY

○ _____
○ _____
○ _____
○ _____
○ _____
○ _____
○ _____
○ _____

23 THURSDAY

○ _____
○ _____
○ _____
○ _____
○ _____
○ _____
○ _____
○ _____

24 FRIDAY

○ _____
○ _____
○ _____
○ _____
○ _____

25 SATURDAY

○
○
○
○
○
○
○
○

26 SUNDAY

○
○
○
○
○
○
○
○

27 MONDAY

○
○
○
○
○
○
○
○

28 TUESDAY

○
○
○
○

January
2025

29 WEDNESDAY

30 THURSDAY

31 FRIDAY

NOTES

01 SATURDAY

○ _____
○ _____
○ _____
○ _____
○ _____
○ _____
○ _____
○ _____

02 SUNDAY

○ _____
○ _____
○ _____
○ _____
○ _____
○ _____
○ _____
○ _____

03 MONDAY

○ _____
○ _____
○ _____
○ _____
○ _____
○ _____
○ _____
○ _____

04 TUESDAY

○ _____
○ _____
○ _____
○ _____
○ _____

February
2025

05 WEDNESDAY

06 THURSDAY

07 FRIDAY

08 SATURDAY

February
2025

09 SUNDAY
- ○
- ○
- ○
- ○
- ○
- ○
- ○
- ○

10 MONDAY
- ○
- ○
- ○
- ○
- ○
- ○
- ○
- ○

11 TUESDAY
- ○
- ○
- ○
- ○
- ○
- ○
- ○
- ○

12 WEDNESDAY
- ○
- ○
- ○
- ○
- ○

13 THURSDAY

- ○
- ○
- ○
- ○
- ○
- ○
- ○
- ○

14 FRIDAY

- ○
- ○
- ○
- ○
- ○
- ○
- ○
- ○

15 SATURDAY

- ○
- ○
- ○
- ○
- ○
- ○
- ○
- ○

16 SUNDAY

- ○
- ○
- ○
- ○
- ○

February
2025

17 MONDAY
- ○
- ○
- ○
- ○
- ○
- ○
- ○
- ○

18 TUESDAY
- ○
- ○
- ○
- ○
- ○
- ○
- ○
- ○

19 WEDNESDAY
- ○
- ○
- ○
- ○
- ○
- ○
- ○
- ○

20 THURSDAY
- ○
- ○
- ○
- ○
- ○

February
2025

21 FRIDAY

○ _____
○ _____
○ _____
○ _____
○ _____
○ _____
○ _____
○ _____

22 SATURDAY

○ _____
○ _____
○ _____
○ _____
○ _____
○ _____
○ _____
○ _____

23 SUNDAY

○ _____
○ _____
○ _____
○ _____
○ _____
○ _____
○ _____
○ _____

24 MONDAY

○ _____
○ _____
○ _____
○ _____
○ _____

25 TUESDAY
○
○
○
○
○
○
○
○

26 WEDNESDAY
○
○
○
○
○
○
○
○

27 THURSDAY
○
○
○
○
○
○
○
○

28 FRIDAY
○
○
○
○
○

01 SATURDAY

02 SUNDAY

03 MONDAY

04 TUESDAY

March
2025

05 WEDNESDAY

○
○
○
○
○
○
○
○

06 THURSDAY

○
○
○
○
○
○
○
○

07 FRIDAY

○
○
○
○
○
○
○
○

08 SATURDAY

○
○
○
○
○

March
2025

09 SUNDAY
- ○
- ○
- ○
- ○
- ○
- ○
- ○
- ○

10 MONDAY
- ○
- ○
- ○
- ○
- ○
- ○
- ○
- ○

11 TUESDAY
- ○
- ○
- ○
- ○
- ○
- ○
- ○
- ○

12 WEDNESDAY
- ○
- ○
- ○
- ○
- ○

13 THURSDAY

○
○
○
○
○
○
○
○

14 FRIDAY

○
○
○
○
○
○
○
○

15 SATURDAY

○
○
○
○
○
○
○
○

16 SUNDAY

○
○
○
○
○

March
2025

17 MONDAY
○
○
○
○
○
○
○
○

18 TUESDAY
○
○
○
○
○
○
○
○

19 WEDNESDAY
○
○
○
○
○
○
○
○

20 THURSDAY
○
○
○
○
○

March
2025

21 FRIDAY

○
○
○
○
○
○
○
○

22 SATURDAY

○
○
○
○
○
○
○
○

23 SUNDAY

○
○
○
○
○
○
○
○

24 MONDAY

○
○
○
○
○

25 TUESDAY

○ _____
○ _____
○ _____
○ _____
○ _____
○ _____
○ _____
○ _____

26 WEDNESDAY

○ _____
○ _____
○ _____
○ _____
○ _____
○ _____
○ _____
○ _____

27 THURSDAY

○ _____
○ _____
○ _____
○ _____
○ _____
○ _____
○ _____
○ _____

28 FRIDAY

○ _____
○ _____
○ _____
○ _____
○ _____

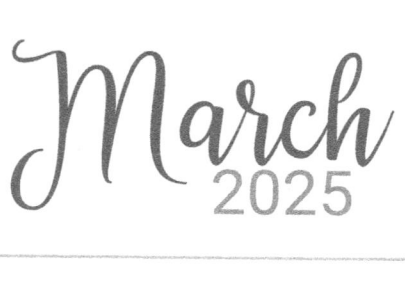

March
2025

29 SATURDAY
○
○
○
○
○
○
○
○

30 SUNDAY
○
○
○
○
○
○
○
○

31 MONDAY
○
○
○
○
○
○
○
○

NOTES

April 2025

01 TUESDAY

○ _____
○ _____
○ _____
○ _____
○ _____
○ _____
○ _____
○ _____

02 WEDNESDAY

○ _____
○ _____
○ _____
○ _____
○ _____
○ _____
○ _____
○ _____

03 THURSDAY

○ _____
○ _____
○ _____
○ _____
○ _____
○ _____
○ _____
○ _____

04 FRIDAY

○ _____
○ _____
○ _____
○ _____
○ _____

April
2025

05 SATURDAY

○ _____
○ _____
○ _____
○ _____
○ _____
○ _____
○ _____

06 SUNDAY

○ _____
○ _____
○ _____
○ _____
○ _____
○ _____
○ _____
○ _____

07 MONDAY

○ _____
○ _____
○ _____
○ _____
○ _____
○ _____
○ _____
○ _____

08 TUESDAY

○ _____
○ _____
○ _____
○ _____
○ _____

April
2025

09 WEDNESDAY

○
○
○
○
○
○
○
○

10 THURSDAY

○
○
○
○
○
○
○
○

11 FRIDAY

○
○
○
○
○
○
○
○

12 SATURDAY

○
○
○
○
○

April
2025

13 SUNDAY

14 MONDAY

15 TUESDAY

16 WEDNESDAY

April
2025

17 THURSDAY

18 FRIDAY

19 SATURDAY

20 SUNDAY

April
2025

21 MONDAY

22 TUESDAY

23 WEDNESDAY

24 THURSDAY

25 FRIDAY

26 SATURDAY

27 SUNDAY

28 MONDAY

April
2025

29 TUESDAY

○ _____
○ _____
○ _____
○ _____
○ _____
○ _____
○ _____
○ _____

30 WEDNESDAY

○ _____
○ _____
○ _____
○ _____
○ _____
○ _____
○ _____
○ _____

NOTES

_____
_____
_____
_____
_____
_____
_____
_____
_____
_____
_____
_____
_____
_____
_____

May
2025

01 THURSDAY

○ _____
○ _____
○ _____
○ _____
○ _____
○ _____
○ _____
○ _____

02 FRIDAY

○ _____
○ _____
○ _____
○ _____
○ _____
○ _____
○ _____
○ _____

03 SATURDAY

○ _____
○ _____
○ _____
○ _____
○ _____
○ _____
○ _____
○ _____

04 SUNDAY

○ _____
○ _____
○ _____
○ _____
○ _____

05 MONDAY

○
○
○
○
○
○
○
○

06 TUESDAY

○
○
○
○
○
○
○
○

07 WEDNESDAY

○
○
○
○
○
○
○
○

08 THURSDAY

○
○
○
○
○

09 FRIDAY

○
○
○
○
○
○
○
○

10 SATURDAY

○
○
○
○
○
○
○
○

11 SUNDAY

○
○
○
○
○
○
○
○

12 MONDAY

○
○
○
○
○

13 TUESDAY

○
○
○
○
○
○
○
○

14 WEDNESDAY

○
○
○
○
○
○
○
○

15 THURSDAY

○
○
○
○
○
○
○
○

16 FRIDAY

○
○
○
○
○

17 SATURDAY

○ _____
○ _____
○ _____
○ _____
○ _____
○ _____
○ _____
○ _____

18 SUNDAY

○ _____
○ _____
○ _____
○ _____
○ _____
○ _____
○ _____
○ _____

19 MONDAY

○ _____
○ _____
○ _____
○ _____
○ _____
○ _____
○ _____
○ _____

20 TUESDAY

○ _____
○ _____
○ _____
○ _____
○ _____

21 WEDNESDAY

○
○
○
○
○
○
○
○

22 THURSDAY

○
○
○
○
○
○
○
○

23 FRIDAY

○
○
○
○
○
○
○
○

24 SATURDAY

○
○
○
○
○

25 SUNDAY

○
○
○
○
○
○
○
○

26 MONDAY

○
○
○
○
○
○
○
○

27 TUESDAY

○
○
○
○
○
○
○
○

28 WEDNESDAY

○
○
○
○
○

May
2025

29 THURSDAY

○ _____
○ _____
○ _____
○ _____
○ _____
○ _____
○ _____
○ _____

30 FRIDAY

○ _____
○ _____
○ _____
○ _____
○ _____
○ _____
○ _____
○ _____

31 SATURDAY

○ _____
○ _____
○ _____
○ _____
○ _____
○ _____
○ _____

NOTES

_____
_____
_____
_____
_____

June
2025

01 SUNDAY

○ _____
○ _____
○ _____
○ _____
○ _____
○ _____
○ _____
○ _____

02 MONDAY

○ _____
○ _____
○ _____
○ _____
○ _____
○ _____
○ _____
○ _____

03 TUESDAY

○ _____
○ _____
○ _____
○ _____
○ _____
○ _____
○ _____
○ _____

04 WEDNESDAY

○ _____
○ _____
○ _____
○ _____
○ _____

June
2025

05 THURSDAY
○
○
○
○
○
○
○
○

06 FRIDAY
○
○
○
○
○
○
○
○

07 SATURDAY
○
○
○
○
○
○
○
○

08 SUNDAY
○
○
○
○

June 2025

09 MONDAY

- ○ _____
- ○ _____
- ○ _____
- ○ _____
- ○ _____
- ○ _____
- ○ _____
- ○ _____

10 TUESDAY

- ○ _____
- ○ _____
- ○ _____
- ○ _____
- ○ _____
- ○ _____
- ○ _____
- ○ _____

11 WEDNESDAY

- ○ _____
- ○ _____
- ○ _____
- ○ _____
- ○ _____
- ○ _____
- ○ _____
- ○ _____

12 THURSDAY

- ○ _____
- ○ _____
- ○ _____
- ○ _____
- ○ _____

June
2025

13 FRIDAY

○ _____
○ _____
○ _____
○ _____
○ _____
○ _____
○ _____
○ _____

14 SATURDAY

○ _____
○ _____
○ _____
○ _____
○ _____
○ _____
○ _____
○ _____

15 SUNDAY

○ _____
○ _____
○ _____
○ _____
○ _____
○ _____
○ _____
○ _____

16 MONDAY

○ _____
○ _____
○ _____
○ _____
○ _____

17 TUESDAY

18 WEDNESDAY

19 THURSDAY

20 FRIDAY

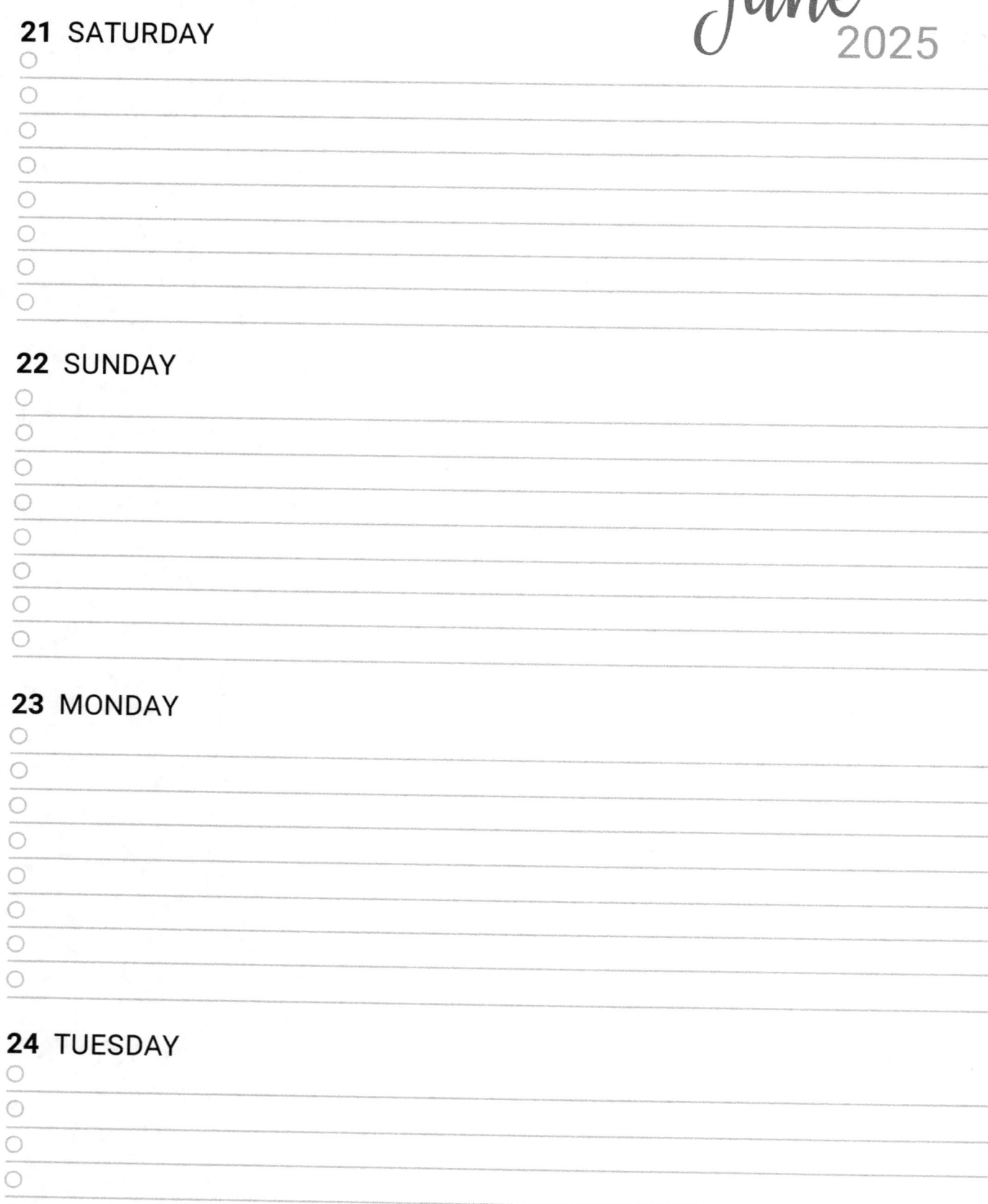

June
2025

21 SATURDAY

22 SUNDAY

23 MONDAY

24 TUESDAY

25 WEDNESDAY

- ○ _____
- ○ _____
- ○ _____
- ○ _____
- ○ _____
- ○ _____
- ○ _____
- ○ _____

26 THURSDAY

- ○ _____
- ○ _____
- ○ _____
- ○ _____
- ○ _____
- ○ _____
- ○ _____
- ○ _____

27 FRIDAY

- ○ _____
- ○ _____
- ○ _____
- ○ _____
- ○ _____
- ○ _____
- ○ _____
- ○ _____

28 SATURDAY

- ○ _____
- ○ _____
- ○ _____
- ○ _____
- ○ _____

June
2025

29 SUNDAY

○ _____
○ _____
○ _____
○ _____
○ _____
○ _____
○ _____
○ _____

30 MONDAY

○ _____
○ _____
○ _____
○ _____
○ _____
○ _____
○ _____
○ _____

NOTES

_____
_____
_____
_____
_____
_____
_____
_____
_____
_____
_____
_____
_____

July
2025

01 TUESDAY

02 WEDNESDAY

03 THURSDAY

04 FRIDAY

July
2025

05 SATURDAY

06 SUNDAY

07 MONDAY

08 TUESDAY

July
2025

09 WEDNESDAY
- ○
- ○
- ○
- ○
- ○
- ○
- ○
- ○

10 THURSDAY
- ○
- ○
- ○
- ○
- ○
- ○
- ○
- ○

11 FRIDAY
- ○
- ○
- ○
- ○
- ○
- ○
- ○
- ○

12 SATURDAY
- ○
- ○
- ○
- ○
- ○

13 SUNDAY

○ _____
○ _____
○ _____
○ _____
○ _____
○ _____
○ _____
○ _____

14 MONDAY

○ _____
○ _____
○ _____
○ _____
○ _____
○ _____
○ _____
○ _____

15 TUESDAY

○ _____
○ _____
○ _____
○ _____
○ _____
○ _____
○ _____
○ _____

16 WEDNESDAY

○ _____
○ _____
○ _____
○ _____
○ _____

17 THURSDAY

○
○
○
○
○
○
○
○

18 FRIDAY

○
○
○
○
○
○
○
○

19 SATURDAY

○
○
○
○
○
○
○
○

20 SUNDAY

○
○
○
○
○

21 MONDAY

○
○
○
○
○
○
○
○

22 TUESDAY

○
○
○
○
○
○
○
○

23 WEDNESDAY

○
○
○
○
○
○
○
○

24 THURSDAY

○
○
○
○
○

25 FRIDAY

26 SATURDAY

27 SUNDAY

28 MONDAY

July
2025

29 TUESDAY

○
○
○
○
○
○
○
○

30 WEDNESDAY

○
○
○
○
○
○
○
○

31 THURSDAY

○
○
○
○
○
○
○
○

NOTES

August
2025

01 FRIDAY

○ _____
○ _____
○ _____
○ _____
○ _____
○ _____
○ _____
○ _____

02 SATURDAY

○ _____
○ _____
○ _____
○ _____
○ _____
○ _____
○ _____
○ _____

03 SUNDAY

○ _____
○ _____
○ _____
○ _____
○ _____
○ _____
○ _____
○ _____

04 MONDAY

○ _____
○ _____
○ _____
○ _____
○ _____

05 TUESDAY

06 WEDNESDAY

07 THURSDAY

08 FRIDAY

August
2025

09 SATURDAY

○ _____
○ _____
○ _____
○ _____
○ _____
○ _____
○ _____
○ _____

10 SUNDAY

○ _____
○ _____
○ _____
○ _____
○ _____
○ _____
○ _____
○ _____

11 MONDAY

○ _____
○ _____
○ _____
○ _____
○ _____
○ _____
○ _____
○ _____

12 TUESDAY

○ _____
○ _____
○ _____
○ _____
○ _____

13 WEDNESDAY

14 THURSDAY

15 FRIDAY

16 SATURDAY

August
2025

17 SUNDAY

18 MONDAY

19 TUESDAY

20 WEDNESDAY

21 THURSDAY

22 FRIDAY

23 SATURDAY

24 SUNDAY

25 MONDAY

○ _____
○ _____
○ _____
○ _____
○ _____
○ _____
○ _____
○ _____

26 TUESDAY

○ _____
○ _____
○ _____
○ _____
○ _____
○ _____
○ _____
○ _____

27 WEDNESDAY

○ _____
○ _____
○ _____
○ _____
○ _____
○ _____
○ _____
○ _____

28 THURSDAY

○ _____
○ _____
○ _____
○ _____
○ _____

August
2025

29 FRIDAY

○ _____
○ _____
○ _____
○ _____
○ _____
○ _____
○ _____
○ _____

30 SATURDAY

○ _____
○ _____
○ _____
○ _____
○ _____
○ _____
○ _____
○ _____

31 SUNDAY

○ _____
○ _____
○ _____
○ _____
○ _____
○ _____
○ _____
○ _____

NOTES

_____
_____
_____
_____

01 MONDAY

○ _____
○ _____
○ _____
○ _____
○ _____
○ _____
○ _____
○ _____

02 TUESDAY

○ _____
○ _____
○ _____
○ _____
○ _____
○ _____
○ _____
○ _____

03 WEDNESDAY

○ _____
○ _____
○ _____
○ _____
○ _____
○ _____
○ _____
○ _____

04 THURSDAY

○ _____
○ _____
○ _____
○ _____
○ _____

05 FRIDAY

○
○
○
○
○
○
○
○

06 SATURDAY

○
○
○
○
○
○
○
○

07 SUNDAY

○
○
○
○
○
○
○
○

08 MONDAY

○
○
○
○
○

09 TUESDAY

○ _____
○ _____
○ _____
○ _____
○ _____
○ _____
○ _____
○ _____

10 WEDNESDAY

○ _____
○ _____
○ _____
○ _____
○ _____
○ _____
○ _____
○ _____

11 THURSDAY

○ _____
○ _____
○ _____
○ _____
○ _____
○ _____
○ _____
○ _____

12 FRIDAY

○ _____
○ _____
○ _____
○ _____
○ _____

September
2025

13 SATURDAY

14 SUNDAY

15 MONDAY

16 TUESDAY

September
2025

17 WEDNESDAY

○ _____
○ _____
○ _____
○ _____
○ _____
○ _____
○ _____
○ _____

18 THURSDAY

○ _____
○ _____
○ _____
○ _____
○ _____
○ _____
○ _____
○ _____

19 FRIDAY

○ _____
○ _____
○ _____
○ _____
○ _____
○ _____
○ _____
○ _____

20 SATURDAY

○ _____
○ _____
○ _____
○ _____
○ _____

21 SUNDAY

- ○
- ○
- ○
- ○
- ○
- ○
- ○
- ○

22 MONDAY

- ○
- ○
- ○
- ○
- ○
- ○
- ○
- ○

23 TUESDAY

- ○
- ○
- ○
- ○
- ○
- ○
- ○
- ○

24 WEDNESDAY

- ○
- ○
- ○
- ○
- ○

25 THURSDAY

○ _____
○ _____
○ _____
○ _____
○ _____
○ _____
○ _____
○ _____

26 FRIDAY

○ _____
○ _____
○ _____
○ _____
○ _____
○ _____
○ _____
○ _____

27 SATURDAY

○ _____
○ _____
○ _____
○ _____
○ _____
○ _____
○ _____
○ _____

28 SUNDAY

○ _____
○ _____
○ _____
○ _____
○ _____

September
2025

29 MONDAY

○ _____
○ _____
○ _____
○ _____
○ _____
○ _____
○ _____
○ _____

30 TUESDAY

○ _____
○ _____
○ _____
○ _____
○ _____
○ _____
○ _____
○ _____

NOTES

_____
_____
_____
_____
_____
_____
_____
_____
_____
_____
_____
_____

October
2025

01 WEDNESDAY
○
○
○
○
○
○
○
○

02 THURSDAY
○
○
○
○
○
○
○
○

03 FRIDAY
○
○
○
○
○
○
○
○

04 SATURDAY
○
○
○
○
○

October
2025

05 SUNDAY

○ _____
○ _____
○ _____
○ _____
○ _____
○ _____
○ _____
○ _____

06 MONDAY

○ _____
○ _____
○ _____
○ _____
○ _____
○ _____
○ _____
○ _____

07 TUESDAY

○ _____
○ _____
○ _____
○ _____
○ _____
○ _____
○ _____
○ _____

08 WEDNESDAY

○ _____
○ _____
○ _____
○ _____
○ _____

October
2025

09 THURSDAY
○
○
○
○
○
○
○
○

10 FRIDAY
○
○
○
○
○
○
○
○

11 SATURDAY
○
○
○
○
○
○
○
○

12 SUNDAY
○
○
○
○
○

October
2025

13 MONDAY

14 TUESDAY

15 WEDNESDAY

16 THURSDAY

October
2025

17 FRIDAY
○ _____
○ _____
○ _____
○ _____
○ _____
○ _____
○ _____
○ _____

18 SATURDAY
○ _____
○ _____
○ _____
○ _____
○ _____
○ _____
○ _____
○ _____

19 SUNDAY
○ _____
○ _____
○ _____
○ _____
○ _____
○ _____
○ _____
○ _____

20 MONDAY
○ _____
○ _____
○ _____
○ _____
○ _____

October

2025

21 TUESDAY

○ _____
○ _____
○ _____
○ _____
○ _____
○ _____
○ _____
○ _____

22 WEDNESDAY

○ _____
○ _____
○ _____
○ _____
○ _____
○ _____
○ _____
○ _____

23 THURSDAY

○ _____
○ _____
○ _____
○ _____
○ _____
○ _____
○ _____
○ _____

24 FRIDAY

○ _____
○ _____
○ _____
○ _____
○ _____

October
2025

25 SATURDAY
- ○
- ○
- ○
- ○
- ○
- ○
- ○
- ○

26 SUNDAY
- ○
- ○
- ○
- ○
- ○
- ○
- ○
- ○

27 MONDAY
- ○
- ○
- ○
- ○
- ○
- ○
- ○
- ○

28 TUESDAY
- ○
- ○
- ○
- ○
- ○

October
2025

29 WEDNESDAY

30 THURSDAY

31 FRIDAY

NOTES

November
2025

01 SATURDAY

-
-
-
-
-
-
-
-

02 SUNDAY

-
-
-
-
-
-
-
-

03 MONDAY

-
-
-
-
-
-
-
-

04 TUESDAY

-
-
-
-

November
2025

05 WEDNESDAY

06 THURSDAY

07 FRIDAY

08 SATURDAY

November
2025

09 SUNDAY

○ _____
○ _____
○ _____
○ _____
○ _____
○ _____
○ _____
○ _____

10 MONDAY

○ _____
○ _____
○ _____
○ _____
○ _____
○ _____
○ _____
○ _____

11 TUESDAY

○ _____
○ _____
○ _____
○ _____
○ _____
○ _____
○ _____
○ _____

12 WEDNESDAY

○ _____
○ _____
○ _____
○ _____
○ _____

November
2025

13 THURSDAY

○ _____
○ _____
○ _____
○ _____
○ _____
○ _____
○ _____
○ _____

14 FRIDAY

○ _____
○ _____
○ _____
○ _____
○ _____
○ _____
○ _____
○ _____

15 SATURDAY

○ _____
○ _____
○ _____
○ _____
○ _____
○ _____
○ _____
○ _____

16 SUNDAY

○ _____
○ _____
○ _____
○ _____
○ _____

November
2025

17 MONDAY

○
○
○
○
○
○
○
○

18 TUESDAY

○
○
○
○
○
○
○
○

19 WEDNESDAY

○
○
○
○
○
○
○
○

20 THURSDAY

○
○
○
○
○

November
2025

21 FRIDAY

22 SATURDAY

23 SUNDAY

24 MONDAY

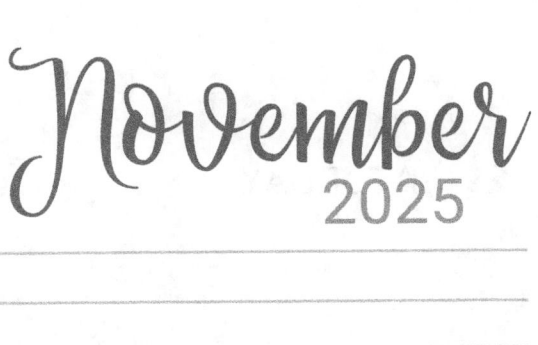

25 TUESDAY

26 WEDNESDAY

27 THURSDAY

28 FRIDAY

November
2025

29 SATURDAY

- ○
- ○
- ○
- ○
- ○
- ○
- ○

30 SUNDAY

- ○
- ○
- ○
- ○
- ○
- ○
- ○
- ○

NOTES

December
2025

01 MONDAY

02 TUESDAY

03 WEDNESDAY

04 THURSDAY

December
2025

05 FRIDAY

06 SATURDAY

07 SUNDAY

08 MONDAY

December
2025

09 TUESDAY

○ _____
○ _____
○ _____
○ _____
○ _____
○ _____
○ _____
○ _____

10 WEDNESDAY

○ _____
○ _____
○ _____
○ _____
○ _____
○ _____
○ _____
○ _____

11 THURSDAY

○ _____
○ _____
○ _____
○ _____
○ _____
○ _____
○ _____
○ _____

12 FRIDAY

○ _____
○ _____
○ _____
○ _____
○ _____

December
2025

13 SATURDAY

14 SUNDAY

15 MONDAY

16 TUESDAY

17 WEDNESDAY

18 THURSDAY

19 FRIDAY

20 SATURDAY

21 SUNDAY

22 MONDAY

23 TUESDAY

24 WEDNESDAY

25 THURSDAY

○ _____
○ _____
○ _____
○ _____
○ _____
○ _____
○ _____
○ _____

26 FRIDAY

○ _____
○ _____
○ _____
○ _____
○ _____
○ _____
○ _____
○ _____

27 SATURDAY

○ _____
○ _____
○ _____
○ _____
○ _____
○ _____
○ _____
○ _____

28 SUNDAY

○ _____
○ _____
○ _____
○ _____
○ _____

December
2025

29 MONDAY

30 TUESDAY

31 WEDNESDAY

NOTES